YOUR KNOWLEDGE HAS VALUE

- We will publish your bachelor's and master's thesis, essays and papers

- Your own eBook and book - sold worldwide in all relevant shops

- Earn money with each sale

Upload your text at www.GRIN.com
and publish for free

Different Causes of Liver Disease in Sri Lanka

Fahim Aslam

Bibliographic information published by the German National Library:

The German National Library lists this publication in the National Bibliography; detailed bibliographic data are available on the Internet at http://dnb.dnb.de.

> ISBN: 9783668632639
> This book is also available as an ebook.

© GRIN Publishing GmbH
Nymphenburger Straße 86
80636 München

All rights reserved

Print and binding: Books on Demand GmbH, Norderstedt, Germany
Printed on acid-free paper from responsible sources.

The present work has been carefully prepared. Nevertheless, authors and publishers do not incur liability for the correctness of information, notes, links and advice as well as any printing errors.

GRIN web shop: https://www.grin.com/document/383789

To identify and examine the different causes of Liver Disease in Sri Lanka

Fahim Aslam

Acknowledgement

I would like to thank my lecturer, Dr. Hashan Kulasena for providing valuable feedbacks and opinions and insights to the project which made it possible for me to complete the project on time. I would also like to thank my seniors and batch members for providing their valuable support for me to complete the project.

Thank you.

Contents

1. Abstract
2. Introduction
3. Materials and Methods
4. Results
5. Discussion
6. Conclusion
7. References

Table 1: The prevalence of alcoholic effects on causing alcoholic disease is examined in the below table with the amount of alcohol consumed and the death rates and alcohol contribution factors for the deaths to occur (WHO, 2014) 7

Table 2: Factors that influence the non-alcoholic fatty liver disease (NAFLD) and the difference between the genders (WHO, 2016). 7

Table 3: Reported Number of Dengue cases over the period of 2010-2017 and the link between the liver disease rate and dengue prevalence (epid.gov.lk, 2017) 8

Table 4: The data set for the three decades for the liver cirrhosis mortality at 95% confidence level is mentioned in the below table, the upward trend of the results are visible for the increasing significance of liver disease (Mokdad et al., 2014) 8

Abstract

Liver disease is one of the main causes for deaths in Sri Lanka, this is the second most common disease causing deaths in hospitals in Sri Lanka after the heart disease. Sri Lanka ranks eighty-nine (89) in the world rankings for liver disease causing 3349 deaths according to the data published by the World Health Organization (WHO) for the 2014 calendar year and an average of 15.28 deaths per hundred thousand. The two most common forms of this disease is non-alcoholic fatty liver disease and alcoholic fatty liver disease. The data collected by the WHO is analyzed and the different causes for the liver disease is identified between the period of 1980-2010, using the different factors responsible for the cause of the disease data is distinguished. Of the data collected and analyzed most causes of liver disease in Sri Lanka is due to the non-alcoholic fatty liver disease (NAFLD) or alcoholic fatty liver disease which leads to severe complications such as renal failure, liver cirrhosis and eventually death.

Introduction

Liver disease is one of the most prominent disease which causes millions of deaths around the world. Sri Lanka has shown vast number of cases develop over the past decades in which the disease has caused several deaths among people, the most common types being non-alcoholic fatty liver disease, alcoholic fatty liver disease, hepatitis A, herbal medicine and toxins and liver disease due to dengue infection (Wijewantha, 2017). According to the data published by the WHO in 2010 for the number of deaths caused by liver disease caused by intake of alcohol is 57% for males among a hundred thousand and 37.7% in females, this percentage highlights the main cause of alcoholic fatty liver disease which leads to the severe form of liver cirrhosis.

The number of deaths from liver cirrhosis from different forms of liver disease has increased severely from 1047 in 1980 to 3415 in 2010, which shows an increase of 330% in the span of 30 years. Sri Lanka had the second highest percentage change for liver cirrhosis mortality from 1990 to 2010 where 80.4% increase was visible (Mokdad,2014). Liver cirrhosis is one of the major problems which has increased the prevalence of increasing mortality. 20% of the world population are suffered by the NAFLD and 10% of the population causes AFLD.

Non-Alcoholic Fatty Liver Disease (NAFLD)

This is one of the common forms of liver disease in Sri Lanka, where extra fat builds up on the liver cells found in the liver. The most common causes of this disease is the people being obese, overweight or diabetes patients. The disease develops in four main stages, hepatic steatosis in which fat begins to build in small amounts, this initially leads to obesity and type 2 diabetes commonly in the patients suffering (Richard and Lingvay, 2011). Sri Lanka has shown high number of reported cases of diabetes over the decades among adults and recently onwards in teenagers as well. According to the DASL (Diabetes Association of Sri Lanka) the diabetes type 2 prevalence has been figured around 20% in Sri Lanka by 2014 and the most recent studies carried by the WHO has shown mortality rate by diabetes in Sri Lanka has risen to 7%, obesity and diabetes contribute to 14.7% in the people (WHO, 2016).

The linkage between the hepatic steatosis and prevalence of diabetes and obesity in Sri Lanka shows a correlation in the results and the studies carried out around the country. This also shows that the linkage between the high mortality rates due to diabetes and rise in cardiovascular diseases such as atherosclerosis also has the fat deposition playing a part in the primary stages of NAFLD and provides evidence that the increase number of causes signifies interests from various parts of the body.

The secondary stage of NAFLD is the NASH (Non-alcoholic steatohepatitis), this is a more severe form of the liver disease caused in which liver inflammation takes place and causes the damage to the liver cells present in the liver, the most common symptoms to lead to these complications are obesity, type 2 diabetes and metabolic syndrome onset inside the body. A study carried out by Kasturiratne on the "influence of non-alcoholic fatty liver disease on the development of diabetes mellitus" found out that out of the 926 patients who were diagnosed with having NAFLD, 676 patients had diabetes mellitus as the most common cause (Kasturiratne, 2013).

The third stage of the disease is known as fibrosis in which the tissues begin to scar the liver tissues and blood vessels present around the liver, the liver function is not severely affected by this since the cells have enough amount of energy and resources for them to function. The final stage and the most severe stage is cirrhosis in which the liver tissue will begin to become absent and lot of fibrosis tissues being to settle on the liver. The structure of the liver begins to change in the process causing the functions to slow down, this could lead to liver failure inside the body.

In Sri Lanka one study was conducted in the Nuwara Eliya district by a team of scientists of the prevalence of NAFLD among the population. 35-64 years old residence took part in the study in which 18% of the population among 403 had NAFLD which was a large percentage among the small population of people. The most common causes identified and examined for these causes were gender, high BMI, high blood glucose and blood pressure. The study identified that the metabolic syndrome and the economic status also played a part in which different types of junk food containing fats and oils contributed in the prevalence of the disease (Pinidiyapathirage et al., 2011).

The number of deaths due to cirrhosis has risen drastically from 1990 to 2010 where 3415 deaths had occurred in which data for NAFLD and AFLD have not been categorized separately.

Alcoholic Fatty Liver Disease (AFLD)

AFLD is another prominent form of liver disease that causes many deaths in Sri Lanka, this liver disease involves the consumption of alcohol which makes the liver accumulate fat in the liver cells present. Consumption of alcohol over a period of time can cause the liver to prevent regeneration of new cells, the chemicals and toxins present in alcohol prevents the filtration process in the liver taking place properly and slows down the regulation of blood sugar and cholesterol levels as the functions of the liver begin to slow down and eventually stop it. The three main types of the disease are fatty liver, which is also known as steatosis just like the primary stage in NAFLD where fat cells begin to accumulate in the liver. According to a research carried out globally the prevalence of steatosis in drinkers were found, 46.4% of heavy drinkers developed steatosis and obesity was common in 94.5% of them (Bellentani, 2017).

The secondary stage is alcoholic hepatitis where the inflammation of the liver takes place and liver cells begin to destroy inside the body. This can be mild as well as in severe forms inside the body depending on the damage done by the intake of the alcohol over a period and the quantity taken. According to the WHO data published for 1961-2010 in Sri Lanka an average of 7.3 liters of alcohol have been consumed by males and 0.3 for females on average which is higher than the average for South-East Asia on 3.5. On average 20.1 liters of alcohol were consumed by the people of both sexes according to the WHO report. (WHO, 2014).

The last stage of the disease is when the liver cirrhosis takes place, the liver is severely damaged during this stage and immediately preventing consumption of alcohol during this stage can cause life expectancy to increase. According to the age standardized death rates (ASDR) and positive correlation factor with alcohol attributable fractions (AAF) for males 37.3 ASDR there is a 57% AAF which shows alcohol contributes most deaths in the severe form of liver cirrhosis in males and in females for 5.3% ASDR there is a 37.7% positive correlation. This data summed up and analyzed Sri Lanka is ranked 4 out of 5 for the number of years lost, higher ranking more the years lost in life (WHO, 2014).

The prevalence of cirrhosis can cause other complications such as accumulation of the fluid in the abdomen and bleeding from the veins, these can lead to problems such respiratory tract infections and renal failure resulting from the liver failure inside the body.

Liver Disease caused by Dengue Infection

Dengue virus infection has been a major threat to Sri Lanka causing severe illness and infection in the country, a reported 95344 cases are filed over the seven-month period. Dengue infection can also be a factor in causing the liver disease in many ways. Dengue disease is spread by the vector carrying the disease, Aedes mosquito.

Dengue virus has four main serotypes in which the disease can be caused DENV 1-4, liver is the main organ which gets involved due to the virus function, the hepatocytes and Kupffer cells found in the liver are the main targets for the dengue infection (Samantha and Sharma, 2015). The DENV binds with the hepatocytes and leads to the cell death eventually by apoptosis, this leads to the cells destroying and necrosis in the liver, the enlargement of the liver takes place and leads to most severe form of liver failure in children than adults.

The levels of AST and ALT increases in the liver after the onset of the dengue fever which then develops into dengue hemorrhagic fever and then finally to dengue shock syndrome in the patient. A research carried out by Samitha Fernando and his team of researchers tested for the liver functional tests values in 55 patients affected by the dengue infection. 22 patients had severe dengue infections and 33 had non severe dengue infections, AST and ALT levels were on the rise in the patients who had severe dengue infection and this showed that the liver had an injury or damage that caused the AST and ALT levels to elevate inside the body. The significance of liver damage causing death in patients cannot be underlining since there is elevation in liver enzymes during the infection illness period. Although no data is available for link between liver disease caused by dengue infection, there is a link in the dengue infection being one of the main contributors of the disease in Sri Lanka with the rising amount of cases and more painkillers taken which increases the toxic substance found inside the body (Fernando et al., 2016).

The main causes of the liver disease in Sri Lanka are as mentioned above, in this article the prevalence of these diseases and the causes among the population is examined and using the data available for the disease the different trends in the change in patterns of the liver disease is identified.

Materials and Methods

Statistical data analysis of the data collected by various journals and WHO on liver disease in Sri Lanka from the period of 1980-2010. The main set of was obtained from a research article published by Ali Mokdad and his team of researchers (Mokdad et al., 2014). The data obtained from the different countries ministry, registration and verbal autopsy on cirrhosis were recorded over a period of three years for the analysis of data for 30 years.

The study was approved by the World Health Organization (WHO) and under the regulation of the International Classification of Diseases (ICD) manual. External data resources for Sri Lanka was obtained using journal data published by Hasitha Wijewantha article on the different types of liver disease (Wijewantha, 2017) and WHO data release for the period of 1961-2010 in Sri Lanka (WHO, 2014).

Measurements

The number of deaths caused by liver cirrhosis is examined in the process of different liver diseases, the most severe form of liver disease also known as cirrhosis causes and the reasons for the death to occur is checked with and against the data on alcohol attribution factors and non-alcoholic attribution factors responsible in the process.

Using resources from researches carried out on the different diseases and the causes for these liver diseases to occur the statistical analysis for each main disease is carried out. The population, gender and type of habits such as alcoholic, non-alcoholic or drug abuse patients are examined and fit into different diseases. Using these data, the most common types of liver disease in Sri Lanka are examined and checked for a period of thirty years and the differential causes for these occurrence is measured.

Results

Table 1: The prevalence of alcoholic effects on causing alcoholic disease is examined in the below table with the amount of alcohol consumed and the death rates and alcohol contribution factors for the deaths to occur (WHO, 2014)

Alcoholic Disease	Males	Females
1. Average Consumption in liters	7.3	0.3
2. Liver Cirrhosis Death Rate (%)	37.3	5.3
3. Alcohol Attributable Factors (%)	57	37.7

Table 2: Factors that influence the non-alcoholic fatty liver disease (NAFLD) and the difference between the genders (WHO, 2016).

Non-Alcoholic Disease	Males	Females
1. Diabetes Prevalence (%)	7.3	8.4
2. Deaths due to diabetes	36.6%	35.7%
3. Deaths due to high blood glucose	59.9%	55.3%
4. Obesity (%)	3.5	10.0

Table 3: Reported Number of Dengue cases over the period of 2010-2017 and the link between the liver disease rate and dengue prevalence (epid.gov.lk, 2017)

Year	Number of Reported Dengue Cases
2010	34188
2011	28473
2012	44461
2013	32063
2014	47502
2015	29777
2016	55150
2017*	97125

*The data for the year 2017 published is for the period of seven months

Table 4: The data set for the three decades for the liver cirrhosis mortality at 95% confidence level is mentioned in the below table, the upward trend of the results are visible for the increasing significance of liver disease (Mokdad et al., 2014)

Year	Number of deaths caused by Liver Cirrhosis
1980	1047
1990	1123
2000	2942
2010	3435

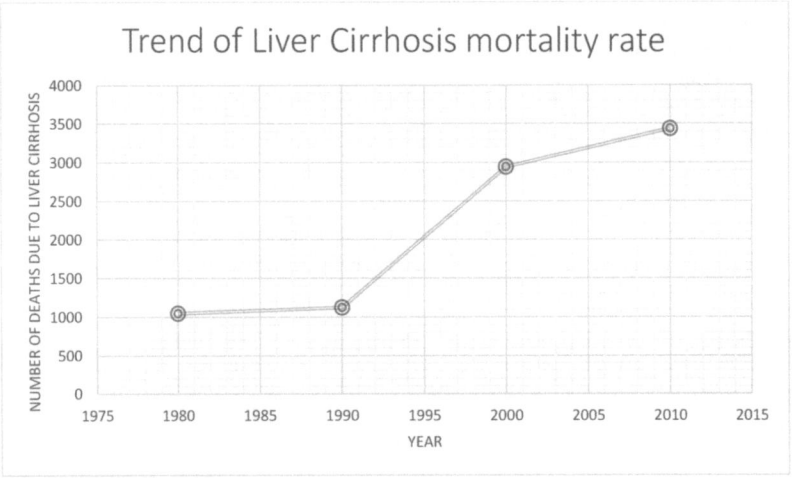

Figure 1: Liver Cirrhosis in Sri Lanka 1980-2010 (Mokdad et al., 2014)

The above figure 1 shows the upward trend of the Liver Cirrhosis mortality in Sri Lanka, the vast peak is observed from 1990 to 2000 where the reported number of deaths have doubled (Table 4: Liver Cirrhosis for three decades). The different causes and the modernization in the country with the trends in the 21st Century is one of the major reasons for the sudden rapid changes.

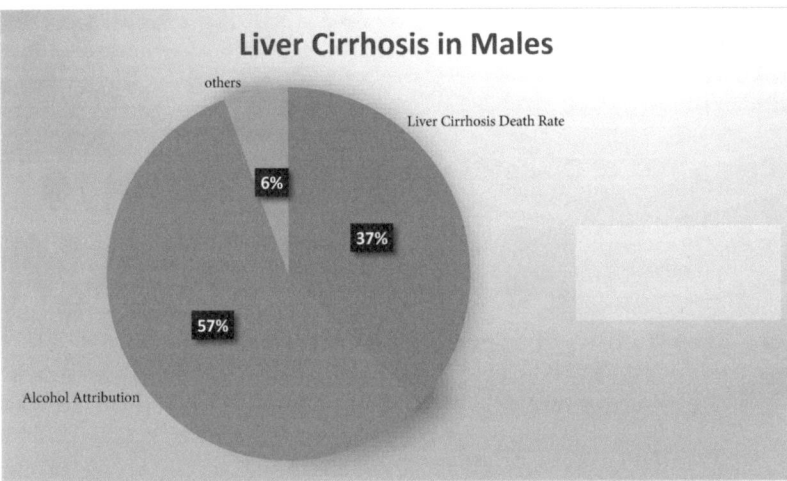

Figure 2: Liver Cirrhosis in Males, the 57% who consumes alcohol results in 37% of the death caused by the alcohol consumed. The other 6% die of reasons not related to alcohol consumption (WHO, 2014).

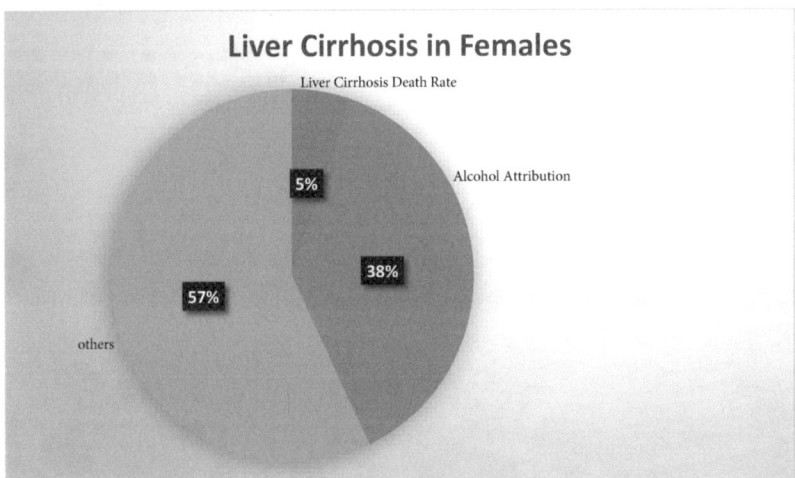

Figure 3: Females show a lower death chances due to alcohol causing liver cirrhosis in which only 5% contributes to the death rate caused by Liver cirrhosis in the patients and 57% are other reasons which cause the death due to Liver cirrhosis (WHO, 2014).

Males and females are both prone to get deaths caused by alcoholic liver disease which goes into the severe form of liver cirrhosis. Females have a relatively lower chance of dying due to liver cirrhosis caused by alcohol as figure 3 mentions the proportion contributing to alcoholic liver cirrhosis that leads to death is only 5% in the females.

This is a complete contradiction to the males since out of the 57% alcohol contribution, 37% is responsible to cause liver cirrhosis that leads to death. The other factors not related to liver cirrhosis deaths due to alcohol is only 6%, this shows that alcohol high consumption my males is a major cause to direct death by liver cirrhosis.

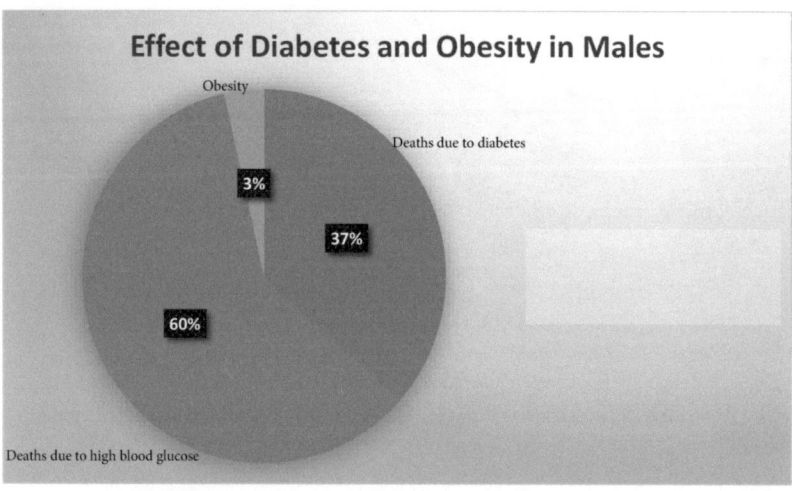

Figure 4: The death rate due to diabetes and high blood glucose levels are 97% in males suffering from the disease in Sri Lanka, 3% of these males end up being obese after severely affected by diabetes. Only 7.3% of the male population have diabetes (WHO,2016).

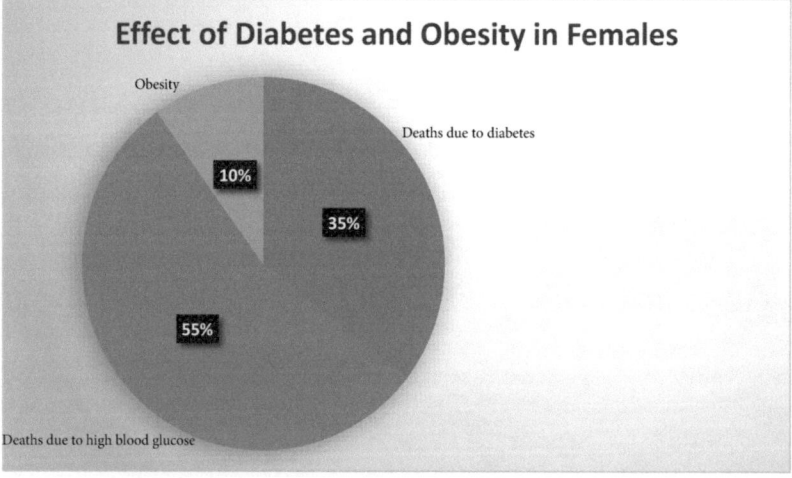

Figure 5: In females the death rates are lower than in males, this changes after menopause in females which causes the glucose imbalance to take place. 8.4% of the population in females (WHO,2016).

Alcohol consumption plays a bigger part in males than in females since the consumption amounts by males are much higher than females (Table 1: Average consumption in males is 7.3 and females is 0.3). This can lead to high chances of the liver cirrhosis causing death in the patients.

Diabetes and obesity also influences the NAFLD chances causing liver disease since one in eleven people in Sri Lanka are affected by the diabetes there are high chances of the body being accumulated with glucose and causing hepatic steatosis. Males have a high death rate due to high blood glucose which is another factor in which glucose is deposited onto the liver. Females have higher chances of being obese than the males in diabetes or high blood glucose conditions inside the body (Table 2: NAFLD and associating factors).

Discussion

NAFLD, Alcoholic liver disease and liver disease caused by dengue infection are three major types of infections that lead to liver cirrhosis, an estimated fifty million people around the world are suffering from chronic liver disease. The consumption of alcohol in individuals results in 9.5% of the liver disease around the globe and prevalence of NASH in NAFLD has grown from 6% to 35% by the end of 2015.

According to the data published by Jurgen Rehm in the article of "Global burden of alcoholic liver diseases" a reported 493300 deaths were caused due to liver cirrhosis in 2010 representing 0.9% of the global deaths. Alcohol induced liver disease was only responsible for 80600 deaths and the rest being other factors such as NASH, NAFLD, hepatitis and other causes (Rehm et al., 2013). Globally the rise of deaths due to liver disease have been under the watch with continuous increase in the number of cases of the disease being spread around the globe. In Asia 15-30% of the general population are suffering from NAFLD according to a research carried out by scientist Wong, (Wong,2012) in which 50% of them are suffering from liver cirrhosis in the body.

In Sri Lanka the rise of liver cirrhosis has increased considerably over the past three decades, (Table 4: Mortality Rate of liver disease over the past three decades). The average percentage for alcohol attributable deaths in South-East Asia region lies at 12.6% for women and 16.2% for men whereas in Sri Lanka males have 57% attributable deaths due to alcohol and females have 38%. This is relatively a high percentage compared to the other South East-Asian countries and shows have an overall growth of 80.4% in the mortality rate due to liver cirrhosis, alcohol related liver disease shows a vast climb due to the high contribution of alcohol in the cause of liver disease (Rehm et al., 2013). Patients suffering from alcoholic liver disease end up losing a lot of years in their life in which the average for South East-Asia lies on 14.6% of the age lost and in Sri Lanka the age lost according to the WHO statistics from 2010 is 19.4% for the population in Sri Lanka (WHO, 2014).

Alcoholic liver disease is evident to be one of the major causes of liver disease in Sri Lanka and has a major contribution in the disease to develop into liver cirrhosis which can lead to death in the patients, this is due to several reasons among Sri Lankan population. Sri Lankan made alcoholic beverages are the most common form of purchasing to be made in the country where people are able to consume more alcohol at a cheaper cost and the quality standards of the alcohol produced in the country have resulted in many deaths due to high toxicity. An example of this is the amount of alcohol consumed in Sri Lanka for the year 2010 where males have an average of 7.3 liters of legal alcohol being consumed according to the WHO report for Sri Lanka for the year 2010, and Sri Lanka ranks number one for the most amount of alcohol consumed liters per capita at 16.79

liters. The most common drink to be consumed is arrack in Sri Lanka where 75.25 liters on average were consumed which was 86% of the major source (Sundaytimes.lk, 2012). This also has influenced in Sri Lanka being the second highest death rate for liver cirrhosis at 55 deaths per hundred thousand only next to Moldova.

The consumption of arrack causes the fat deposition to increase over the time in the body and results in the toxicity released by the arrack produced, the liver is unable to deal with the toxic substance eventually leading to the liver failing. Arrack consumption is more prominent in low income families of the urban region where they consume arrack at a larger scale (Katulanda et al., 2014). The main reasons for consuming alcohol in Sri Lanka were for tiredness, joy, sorrow and for courage, this makes people drink high amounts of alcohol making the body fight with high toxic levels in the liver. The liver is unable to neutralize and excrete these toxic products inside the body resulting in the disturbance in both the mental and physical health, due to alcoholism many road traffic accidents, violence, sexual abuse and other diseases also happen.

The main reasons for the country of 20 million people to be ranked number one in the most amount of alcohol consumed is mainly the income levels of the people where most people with lower income prefer intake of high toxic drinks such as toddy and arrack which are cheap to purchase and the levels of toxicity is also high. The other reason is no health policy is developed in the country for the drinkers and no national plans are available in the country to minimize the alcohol consumption (WHO, 2014). There is no restriction for how many hours the stores selling alcohol can function in the country unlike countries like USA where policies are designed to increase and reduce sale hours during different times of the seasons (Hahn et al., 2010). There is no restriction in selling alcohol for the people who have been already intoxicated and under severe disease control this makes the people who have been already intoxicated to be able to consume more alcohol over the period of time in the country without any ban imposed on them. No regulations and health warnings are placed on the alcohol being sold and no advertising is carried out on the sale of alcohol in the country allowing people to be less aware of the disease caused in the country and the implications caused by it (WHO, 2014).

NAFLD on the other hand has been one of the major issues surrounding the world since 20% of the world population are suffering from NAFLD and most of these people are affected with obesity and diabetes in common. Sri Lanka have a high rate of people with diabetes and obesity making them prone to get NAFLD. No proper study among a large population is not carried out to analyze the effect of diabetes and obesity linked up with NAFLD. A study carried out by Pinidiyapathirage in 2011 identified the NAFLD among the population living in Nuwara Eliya where the population was checked for being obese and having a high BMI in them causing the NAFLD to be present in 18% in the checked population (Pinidiyapathirage et al., 2011).

Sri Lanka has a high rate of prevalence of diabetes in the world, where Sri Lanka ranks 31 out of the world for the number of death rates per hundred thousand where Sri Lanka stands at 47.87. According to a research carried out by Katulanda, the Sri Lankan Tamil and Moor ethnicity had the highest rate of diabetes present in them (Katulanda et al., 2011). Diabetes causes 7% of the mortality among Sri Lankan population and causes nearly 10000 deaths every year. The patients suffering or recovered from diabetes are either with high blood glucose, which results in around 15000 deaths (Table 2: Factors influencing non-alcoholic fatty liver disease). This population are obese, overweight, physically inactive or diabetes prone patients.

The link between NAFLD and diabetes patients are that patients with NAFLD have extra fat deposits on the liver cells causing it to go into NASH and eventually cirrhosis in the patients. Sri

Lanka has shown an inclined rate of liver cirrhosis ranking second in the world behind Moldova where the alcoholic factors contribute to around 50% of the causes for the disease. The other 50% lies among the other liver diseases and NAFLD is one them. In Sri Lanka the urban population suffers from 32.6% NAFLD and 18% suffers in the rural area (Kodisinghe and Niriella, 2015). According to the World Gastroenterology Organization, NAFLD and NASH are the main causes for advanced liver disease and mortality around the world (Sherif, 2016) and the main causes for this to happen have been identified as diabetes, obesity and metabolic syndrome disorders.

Adolescents are more viable to get these diseases and in Sri Lanka in a survey carried out this was proven when out of the 508 adolescents involved 44 of them had NAFLD with obesity, high waist and elevated blood pressure visible (Rajindrajith et al., 2015). Sri Lanka have a high proportion of obese people and diabetes people where around 6.8% of the Sri Lankan suffer from obesity and 26.1% are overweight. No health polices and strategies available in the country has made the prevalence of the disease to be at its peak in Sri Lanka since the number of diabetes cases have increased gradually and the number of cases reporting from NAFLD also have increased since the rate of liver cirrhosis have increased significantly. No diabetes registry is maintained to identify the number of deaths caused due to various reasons in the people and no proper medications are available on hand as primary facilities for the people suffering from the disease to minimize the effect such as the availability of insulin as the primary healthcare facility (WHO, 2016).

The growth of NAFLD cases and NASH cases are also due to the food intake and the amount of physical activity taken by the people, out of the 20 million Sri Lankan population present 23.7% of them are physically inactive causing diseases such as artherosceloris and respiratory tract infections to develop. The food containing high sugar levels can result in high blood glucose in the patients and the chances of developing diabetes in them. The modern era of fast food influences most the causes of NAFLD due to the food containing high levels of fat and oils and deposits inside the body in excess causing various diseases to be formed. According to a journal published by Yasutake the dietary habits, lifestyle and the therapeutic approach carried out for the NAFLD patients influences the significance of the stages development of the NAFLD. Low calorie diet and exercise are examples of things that needs to be followed to prevent from being affected by NAFLD (Yasutake et al., 2014).

The high rates of death due to cardiovascular diseases also can be influenced by the NAFLD since it could lead to diabetes as well in the people. The type 2 diabetes and coronary heart disease are the most common forms of the diseases to be formed and this does not contain data in Sri Lanka to analyze and distinguish between cases formed due to NAFLD. Lack of data available to analyze the different causes produced by the NAFLD and the complications that can rise due to it cannot be discussed due to limited information available on the disease in Sri Lanka.

The most common causes of the liver disease in Sri Lanka are most likely to be NAFLD and AFLD in the patients who die due to liver cirrhosis since most the factors and data collected are within the parameters of the results obtained. There are other diseases and causes as well which can cause the liver cirrhosis to take place such as the dengue infection which can lead to liver damage where the viral toxicity and the apoptosis of hepatocytes in the liver. With Sri Lanka having increased cases of dengue fever cases the chances of liver cirrhosis also can increase and become significant in the near future if no treatment for dengue is discovered (Samantha and Sharma, 2015). Another disease that can cause the increased chances of liver cirrhosis is viral hepatitis which can cause viral replication that can result in the hepatocytes to produce more proteins of the viruses and cause severe damage to the liver (Nakamoto and Kaneko, 2003). Hepatocellular carcinoma is another disease which is not that evident of causing liver disease in Sri Lanka but is one of the most common forms of cancer present in the world. In Sri Lanka a 105-reported cases were found in 2011 in the North Colombo Liver Unit, although the disease data is not available there are chances of the data of HCC to be presented in the coming years causing liver cirrhosis (Siriwardana et al., 2013).

Liver cirrhosis is a growing cause present in Sri Lanka and has shown a significant rise from 1990 to 2010 where the reported number of deaths have increased by 80.4% in the country and the data not measured properly by the Sri Lankan Health databases. There are several limitations regarding the type of liver disease causing liver cirrhosis since the data collected was analyzed using various sources in which the liver disease was examined and analyzed. The cause of death due to liver cirrhosis and the main reasons behind the cause of it happening cannot be identified accurately since various diseases resembles the cirrhosis to occur such as NAFLD and AFLD the exact cause for the cirrhosis of the patients cannot be known without the data for the different diseases. No documentation of the results and the liver disease after 2014 found to be available where only small researches were carried out over time to identify the widespread of the disease present therefore the prediction of the liver disease from 2014 is unknown. Various other reasons excluded in which liver cirrhosis can take place such as hepatitis and HCC are the two most common forms present in the world in which can cause liver cirrhosis at a large scale. Data obtained based on hospital records of people registered for the treatment and the people with the severe forms of liver disease only, this could limit the findings for the preliminary stages of NAFLD and AFLD since the records are only available for the cirrhosis state.

Conclusion

In study the project identifies the different types of liver disease and the main factors that influences them. The most common types of liver diseases are identified as NAFLD and AFLD which causes the most severe form of liver cirrhosis in patients. The main factors that influences NAFLD is diabetes and obesity and for AFLD is the consumption of alcohol. Males are the most affected in AFLD and females and males have almost equal effects in NAFLD. To reduce the number of liver cirrhosis cases the NAFLD and AFLD needs to be diagnosed at earliest stages of the disease and record the data to ensure that the other population does not go into the same trend of suffering. Sri Lanka is ranked second for the most number of deaths due to liver cirrhosis and ranked number one in the consumption of alcoholic beverages in the country, this has been progressing over the years where 80.4% increased deaths from 1990 to 2010 was observed. Future action needs to be taken to ensure the alcohol consumption is reduced and health policies are implemented to reduce cases of diabetes and alcoholism in Sri Lanka.

Abbreviations – NAFLD, non-alcoholic fatty liver disease; AFLD, alcoholic fatty liver disease; WHO, World Health Organization; BMI, body mass index; DASL, Diabetes Association of Sri Lanka; NASH, non-alcoholic steatohepatitis; ASDR, age standardized death rates; AAF, alcohol attributable fractions; DENV, dengue virus; AST, aspartate transaminase; ALT, alanine transaminase; ICD, International Classification of Diseases; HCC, hepatocellular carcinoma.

References

1. Abeysekera, R., Illangasekera, U., Jayalath, T., Sandeepana, A. and Kularatne, S. (2013). Successful use of intravenous N-acetylcysteine in dengue haemorrhagic fever with acute liver failure. *Ceylon Medical Journal*, 57(4).
2. AlKhater, S. (2015). Paediatric non-alcoholic fatty liver disease: an overview. *Obesity Reviews*, 16(5), pp.393-405.
3. Asia, S. (2016). Public health round-up. *Bulletin of the World Health Organization*, 94(2), pp.80-81.
4. Bellentani, S. (2017). The epidemiology of non-alcoholic fatty liver disease. *Liver International*, 37, pp.81-84.
5. Bentzen, J. and Smith, V. (2011). Alcohol Consumption and Liver Cirrhosis Mortality: New Evidence from a Panel Data Analysis for Sixteen European Countries. *Journal of Wine Economics*, 6(01), pp.67-82.
6. Eshraghian, A. (2017). High prevalence of nonalcoholic fatty liver disease in the middle east: Lifestyle and dietary habits. *Hepatology*, 65(3), pp.1077-1077.
7. Fernando, S., Wijewickrama, A., Gomes, L., Punchihewa, C., Madusanka, S., Dissanayake, H., Jeewandara, C., Peiris, H., Ogg, G. and Malavige, G. (2016). Patterns and causes of liver involvement in acute dengue infection. *BMC Infectious Diseases*, 16(1).
8. Hahn, R., Kuzara, J., Elder, R., Brewer, R., Chattopadhyay, S., Fielding, J., Naimi, T., Toomey, T., Middleton, J. and Lawrence, B. (2010). Effectiveness of Policies Restricting Hours of Alcohol Sales in Preventing Excessive Alcohol Consumption and Related Harms. *American Journal of Preventive Medicine*, 39(6), pp.590-604.
9. Jayasinghe, S. (2013). Illness and social protection: an agenda for action in Sri Lanka. *Sri Lanka Journal of Social Sciences*, 33(1-2).
10. Katulanda, P., Jayawardena, M., Sheriff, M., Constantine, G. and Matthews, D. (2010). Prevalence of overweight and obesity in Sri Lankan adults. *Obesity Reviews*, 11(11), pp.751-756.
11. Katulanda, P., Jayawardena, M., Sheriff, M., Constantine, G. and Matthews, D. (2010). Prevalence of overweight and obesity in Sri Lankan adults. *Obesity Reviews*, 11(11), pp.751-756.
12. Katulanda, P., Ranasinghe, C., Rathnapala, A., Karunaratne, N., Sheriff, R. and Matthews, D. (2014). Prevalence, patterns and correlates of alcohol consumption and its' association with tobacco smoking among Sri Lankan adults: a cross-sectional study. *BMC Public Health*, 14(1).
13. Katulanda, P., Rathnapala, D., Sheriff, R. and Matthews, D. (2012). Province and ethnic specific prevalence of diabetes among Sri Lankan adults. *Sri Lanka Journal of Diabetes Endocrinology and Metabolism*, 1(1).
14. Kodisinghe, S. and Niriella, M. (2015). Evidence-based management of non-alcoholic fatty liver disease. *Sri Lanka Journal of Diabetes Endocrinology and Metabolism*, 5(1), p.28.
15. Mokdad, A., Lopez, A., Shahraz, S., Lozano, R., Mokdad, A., Stanaway, J., Murray, C. and Naghavi, M. (2014). Liver cirrhosis mortality in 187 countries between 1980 and 2010: a systematic analysis. *BMC Medicine*, 12(1).

16. Niriella, M., Pathmeswaran, A., De Silva, S., Kasturiratna, A., Perera, R., Subasinghe, C., Kodisinghe, K., Piyaratna, C., Rishikesawan, V., Dassanayaka, A., De Silva, A., Wickramasinghe, R., Takeuchi, F., Kato, N. and de Silva, H. (2017). Incidence and risk factors for non-alcoholic fatty liver disease: A 7-year follow-up study among urban, adult Sri Lankans. *Liver International*.
17. Pinidiyapathirage, M., Dassanayake, A., Rajindrajith, S., Kalubowila, U., Kato, N., Wickremasinghe, A. and de Silva, H. (2011). Non-alcoholic fatty liver disease in a rural, physically active, low income population in Sri Lanka. *BMC Research Notes*, 4(1), p.513.
18. Ramstedt, M. (2001). Per capita alcohol consumption and liver cirrhosis mortality in 14 European countries. *Addiction*, 96(1s1), pp.19-33.
19. Rehm, J., Samokhvalov, A. and Shield, K. (2013). Global burden of alcoholic liver diseases. *Journal of Hepatology*, 59(1), pp.160-168.
20. Saliba, F. and Samuel, D. (2013). Acute liver failure: Current trends. *Journal of Hepatology*, 59(1), pp.6-8.
21. Sherif, Z., Saeed, A., Ghavimi, S., Nouraie, S., Laiyemo, A., Brim, H. and Ashktorab, H. (2016). Global Epidemiology of Nonalcoholic Fatty Liver Disease and Perspectives on US Minority Populations. *Digestive Diseases and Sciences*, 61(5), pp.1214-1225.
22. Siriwardana, R., Liyanage, C. and Gunethileke, M. (2013). Hepatocellular carcinoma in Sri Lanka - where do we stand? *Sri Lanka Journal of Surgery*, 31(2).
23. Sugimoto, K. and Takei, Y. (2016). Pathogenesis of alcoholic liver disease. *Hepatology Research*, 47(1), pp.70-79.
24. Sundaytimes.lk, (2012). We are a nation of boozers. [online]. Available at : https://www.sundaytimes.lk/120026/Plus/plus_10.html [Accessed July 12. 2017]
25. The ILTS 19thAnnual International Congress. (2013). *Liver Transplantation*, 19, pp. S1-S334.
26. TUYAMA, A. and CHANG, C. (2012). Non-alcoholic fatty liver disease. *Journal of Diabetes*, 4(3), pp.266-280.
27. WHO, J. (2014). Alcohol policy scores: data and analysis. *Bulletin of the World Health Organization*, 93(10), pp.740A-740B.
28. WHO, S. (2015). Noncommunicable diseases: stepping up the fight. *Bulletin of the World Health Organization*, 93(1), pp.9-10.
29. WHO, S. (2016). The mysteries of type 2 diabetes in developing countries. *Bulletin of the World Health Organization*, 94(4), pp.241-242.
30. Wijayagunawardane, M., Wijerathne, C. and Herath, C. (2015). Indigenous Herbal Recipes for Treatment of Liver Cirrhosis. *Procedia Chemistry*, 14, pp.270-276.
31. Wijewantha, H. (2017). Liver Disease in Sri Lanka. *Euroasian Journal of Hepato-Gastroenterology*, 7(1), pp.78-81.
32. Wlodzimirow, K., Abu-Hanna, A. and Chamuleau, R. (2013). Acute-on-chronic liver failure – Its definition remains unclear. *Journal of Hepatology*, 59(1), pp.190-191.
33. Wong, V. (2012). Nonalcoholic fatty liver disease in Asia: A story of growth. *Journal of Gastroenterology and Hepatology*, 28(1), pp.18-23.
34. Yasutake, K. (2014). Dietary habits and behaviors associated with nonalcoholic fatty liver disease. *World Journal of Gastroenterology*, 20(7), p.1756.
35. Zelber-Sagi, S. (2011). Nutrition and physical activity in NAFLD: An overview of the epidemiological evidence. *World Journal of Gastroenterology*, 17(29), p.3377.

YOUR KNOWLEDGE HAS VALUE

- We will publish your bachelor's and master's thesis, essays and papers

- Your own eBook and book - sold worldwide in all relevant shops

- Earn money with each sale

Upload your text at www.GRIN.com
and publish for free